T0269206

Salience Network
of the Human Brain

Salience Network of the Human Brain

Lucina Q. Uddin

ELSEVIER

AMSTERDAM • BOSTON • HEIDELBERG • LONDON
NEW YORK • OXFORD • PARIS • SAN DIEGO
SAN FRANCISCO • SINGAPORE • SYDNEY • TOKYO

Academic Press is an imprint of Elsevier

Academic Press is an imprint of Elsevier
125 London Wall, London EC2Y 5AS, United Kingdom
525 B Street, Suite 1800, San Diego, CA 92101-4495, United States
50 Hampshire Street, 5th Floor, Cambridge, MA 02139, United States
The Boulevard, Langford Lane, Kidlington, Oxford OX5 1GB, United Kingdom

British Library Cataloguing-in-Publication Data
A catalogue record for this book is available from the British Library

Library of Congress Cataloging-in-Publication Data
A catalog record for this book is available from the Library of Congress

ISBN: 978-0-12-804593-0

For Information on all Academic Press publications
visit our website at https://www.elsevier.com

  Working together
to grow libraries in
developing countries

www.elsevier.com • www.bookaid.org

Publisher: Mara Conner
Acquisition Editor: April Farr
Editorial Project Manager: Timothy Bennett
Production Project Manager: Sruthi Satheesh
Cover Designer: Matthew Limbert

Typeset by MPS Limited, Chennai, India

DEDICATION

For my parents

CONTENTS

LIST OF FIGURES

ACKNOWLEDGMENT

Research funded by the National Institute of Mental Health (R01MH107549).

CHAPTER 1

What Is Salience?

At any given moment, our senses are bombarded with information from a variety of sources. We know that flashing lights and sirens signal emergencies, that the presence of dangerous predators can keep us vigilant, and emotional memories might arise at the slightest provocation. Our brains must continuously parse this abundance of information to allow us to successfully navigate the environment. In order to do this, our nervous system must somehow determine what is critical to direct attention to, and what can safely be ignored. Things that are more salient naturally attract more attention. But what exactly do we mean by "salience"?

When we use the term "salience" in daily life, we often mean to convey the Merriam-Webster dictionary definition of the term: "the quality or state of being salient," where salient is "very important or noticeable." Objects, ideas, or events that are "very important or noticeable" have a privileged status in that they attract significant attention and can occupy a disproportionate amount of space in our imaginations. The term salience has been used in multiple domains of psychology and neuroscience to convey this concept of importance and noticeability.

In perception research, and specifically in studies of the visual system, the term salience is typically used to describe aspects of an item that make it stand out relative to its neighbors. For example, the well-known "pop out" effect during visual search (Treisman, 1998) describes the phenomenon that it is easier to identify a target that differs from distractors on one feature than on several features. Certain stimuli are salient by virtue of ease of discrimination from the surroundings (Fig. 1). This type of automatic and effortless salience detection enables rapid direction of visual attention.

The concept of a "saliency map" (Koch & Ullman, 1985), derived from empirical work in visual search, is the idea that a two-dimensional map encodes the saliency of objects in the visual environment.

Salience Network of the Human Brain. DOI: http://dx.doi.org/10.1016/B978-0-12-804593-0.00001-1

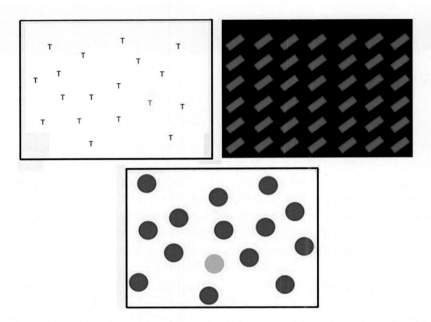

*Figure 1 **Salience in visual search.** Some stimuli are salient by virtue of the "pop out" effect (Treisman, 1998). It is easier to identify a target that differs from distractors on one feature (e.g., color) than on several features. Certain stimuli—the red T, the green bar, and the pink circle—are salient by virtue of ease of discrimination from their surroundings.*

Computational models of this process are based on the idea that early visual features such as color and intensity are computed based on retinal input and activity from these feature maps is combined at each location, giving rise to a topographic saliency map. The "winner-take-all" system subsequently detects the most salient location and directs attention toward it (Itti & Koch, 2000). This type of model is limited to the bottom-up, stimulus-driven control of attention. However, contextual factors and previous experience can also contribute to perceptions of what is salient.

Most theories of salience detection and attention adopt a two-component framework incorporating both bottom-up processes such as the ones just described, as well as top-down contextual influences (Itti & Koch, 2001). In an influential model of attention, Corbetta and colleagues describe two partially segregated networks: a ventral-attention network responding to external environmental stimuli, and a dorsal-attention network responsible for goal-directed, top-down processing (Corbetta & Shulman, 2002). Chapters 2 and 3, Anatomy of the Salience Network and Functions of the Salience Network, will further elaborate on the anatomy and function of these networks as they relate to salience processing.

In areas outside of vision research and computational modeling of attention, salience is often described using a different emphasis, namely that of personal relevance. In such cases, stimuli and/or events that are meaningful or emotionally provocative are termed as salient. This conceptualization of salience as a relatively high-level cognitive process can be seen, e.g., in the literature on emotional learning (Dunsmoor, Murty, Davachi, & Phelps, 2015). In the area of memory research, saliency is thought to influence the likelihood that an event or object will be remembered; objects can be salient because of their meaning or semantic relationship with other objects. Some have proposed that perceptual- (based on low-level sensory features) and semantics-related (based on prior knowledge) salience affect encoding and memory representation in different ways, acting through dorsal and ventral neural systems, respectively (Santangelo, 2015).

In the addiction literature, the term incentive salience or motivational salience refers to the "wanting" that accompanies most addictions, and is thought to be modulated by dopamine (Tibboel, De Houwer, & Van Bockstaele, 2015). As with the definitions of salience we have discussed, the term salience again highlights the property of standing out or being attention-grabbing, in this case to the point of pathology.

As we will review throughout the next several chapters, the concept of salience is ubiquitous in psychology and neuroscience; thus a great deal of research has been devoted to understanding the mechanisms by which salience detection occurs in the brain and the anatomical structures that support it.

REFERENCES

Corbetta, M., & Shulman, G. L. (2002). Control of goal-directed and stimulus-driven attention in the brain. *Nature Reviews. Neuroscience, 3*(3), 201–215.

Dunsmoor, J. E., Murty, V. P., Davachi, L., & Phelps, E. A. (2015). Emotional learning selectively and retroactively strengthens memories for related events. *Nature, 520*(7547), 345–348.

Itti, L., & Koch, C. (2000). A saliency-based search mechanism for overt and covert shifts of visual attention. *Vision Research, 40*(10–12), 1489–1506.

Itti, L., & Koch, C. (2001). Computational modelling of visual attention. *Nature Reviews. Neuroscience, 2*(3), 194–203.

Koch, C., & Ullman, S. (1985). Shifts in selective visual attention: Towards the underlying neural circuitry. *Human Neurobiology, 4*(4), 219–227.

Santangelo, V. (2015). Forced to remember: When memory is biased by salient information. *Behavioural Brain Research, 283*, 1–10.

Tibboel, H., De Houwer, J., & Van Bockstaele, B. (2015). Implicit measures of "wanting" and "liking" in humans. *Neuroscience and Biobehavioral Reviews, 57*, 350–364.

Treisman, A. (1998). Feature binding, attention and object perception. *Philosophical Transactions of the Royal Society of London. Series B, Biological Sciences, 353*(1373), 1295–1306.

CHAPTER 2

Anatomy of the Salience Network

The view that the brain functions as a vast, interconnected network currently dominates the cognitive neuroscience landscape (Pessoa, 2014). The older idea that the brain comprises discrete "modules" for performance of specific functions has given way to more recent models that emphasize dynamics, connectivity, and large-scale brain networks underlying cognition (Barrett & Satpute, 2013; Bressler & Menon, 2010). Resting state functional magnetic resonance imaging (rsfMRI), first used by Biswal and colleagues to examine functional connectivity within the motor system in the absence of task performance (Biswal, Yetkin, Haughton, & Hyde, 1995), has emerged as a powerful tool for discovering the intrinsic architecture and network structure of the human brain.

Brain areas that exhibit strong functional connectivity (e.g., temporal correlations in signal) (Friston, 1994) are thought to form large-scale brain networks that are reproducible across individuals (Damoiseaux et al., 2006) and relatively stable (Shehzad et al., 2009). The most well studied of these brain networks is the default mode network (Greicius, Krasnow, Reiss, & Menon, 2003), first noted for its high metabolic activity during baseline resting states (Raichle et al., 2001) and subsequently linked with prospective memory (Buckner & Carroll, 2007), self-related, and social cognitive processes (Uddin, Iacoboni, Lange, & Keenan, 2007). Large-scale brain networks corresponding to attention (Fox, Corbetta, Snyder, Vincent, & Raichle, 2006) and numerous other cognitive processes have also been identified using rsfMRI. These networks bear strong resemblance to those actively engaged during a variety of cognitive paradigms (Smith et al., 2009).

In the context of modern functional neuroimaging, the first use of the term "salience network" in the literature can be traced to a seminal paper by Seeley and colleagues, in which the now widely used nomenclature of "intrinsic connectivity networks" (ICNs) was also introduced. Using both

Salience Network of the Human Brain. DOI: http://dx.doi.org/10.1016/B978-0-12-804593-0.00002-3

region-of-interest (ROI) and independent-component analysis, Seeley and colleagues demonstrated the existence of an ICN including frontoinsular (FIC)/anterior insular (AI) cortices, dorsal anterior cingulate cortex (dACC), dorsomedial thalamus, hypothalamus, periaqueductal gray, sublenticular extended amygdala, substantia nigra/ventral tegmental area, and temporal pole (Seeley et al., 2007). Prior to this work, the term "salience network" did not appear in descriptions of human neuroimaging studies examining the neural basis of salience processing, although networks with some overlapping nodes had been previously described.

Subsequent studies delineated the anatomy of the salience network in greater detail. As reviewed in Seeley et al. (2012) and Uddin (2015), salience network communication with subcortical regions enables integration of interoceptive and visceromotor signals, which can be used to guide behavior. Ascending inputs from visceroautonomic sensors are integrated in the FIC within the salience network. Interoceptive signals travel via the vagus nerve, through autonomic afferent nuclei and the thalamus, and via the dorsal posterior insular cortex and mid-insula (midINS) to the FIC. The cortical nodes of the salience network integrate these ascending signals to coordinate other large-scale networks in the cortex. The salience network also sends information to visceromotor central pattern generators, which send signals to autonomic efferent nuclei such as the nucleus of the solitary tract and the dorsal motor nucleus of the vagus nerve. These signals then travel to the intermediolateral cell column of the spinal cord, which drives visceroautonomic responses to salient stimuli (Fig. 2).

The anatomy of the insular cortex is of particular importance in enabling salience network function. Located deep within the lateral sulcus of the brain, the insula has traditionally been described as paralimbic or limbic integration cortex (Augustine, 1996). Structural connections delineated in the macaque suggest that the insula communicates with amygdala, orbitofrontal cortex, olfactory cortex, anterior cingulate cortex (ACC), and superior temporal sulcus (Mufson & Mesulam, 1982), thus positioning it at the crossroads of affective, homeostatic, and cognitive systems in the brain.

The ACC comprises subareas with distinct anatomical and functional properties. The observation that dense interconnections exist between the ACC and motor cortices, lateral prefrontal regions, and thalamic and brainstem nuclei led to the suggestion that the region

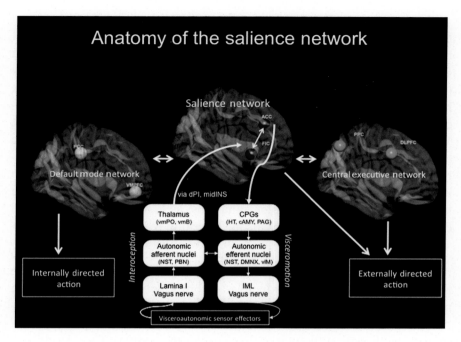

*Figure 2 **Anatomy of the salience network.** Ascending inputs communicating information about the condition of the body are integrated in the frontoinsular cortices (FIC) within the salience network. Interoceptive signals travel through the vagus nerve through autonomic afferent nuclei (NST, nucleus of the solitary tract; PBN, parabrachial nucleus) and the thalamus (vmPO, ventromedial nucleus of the thalamus, posterior; vmB, ventromedial nucleus of the thalamus, basal) onto the FIC via dorsal posterior insula (dPI) and mid-insula (midINS). The salience network communicates with visceromotor central pattern generators (CPGs: HT, hypothalamus; cAMY, central nucleus of the amygdala; PAG, periaqueductal gray) that convey signals to autonomic efferent nuclei (NST: DMNX, dorsal motor nucleus of the vagus nerve) and the vagus nerve (IML, intermediolateral cell column). Salient signals are integrated in the anterior insular cortices. Signals from the anterior insula causally influence the default mode network (DMN, yellow; key nodes in PCC, posterior cingulate cortex; VMPFC, ventromedial prefrontal cortex) and central executive network (CEN, green; key nodes in PPC, posterior parietal cortex; DLPFC, dorsolateral prefrontal cortex). These pathways for communication between the insula and brain regions for interoception and visceromotion allow integration of salient signals to guide behavior (Uddin, 2015).*

may serve an integrative function to translate intentions to actions (Paus, Castro-Alamancos, & Petrides, 2001). The dACC is considered to be a major cortical node of the salience network. Based on structural connectivity-based parcellation, the dACC can be subdivided into three distinct regions (Beckmann, Johansen-Berg, & Rushworth, 2009). As functional connectivity studies demonstrate strong connections between anterior insula and anterior and posterior mid-cingulate cortex (Taylor, Seminowicz, & Davis, 2009), the most likely ACC subdivision to participate in the salience network is one that is centered midway along the cingulate gyrus.

An interesting structural feature of a salience network is that its two prominent cortical nodes, the AI and ACC, contain a special type of neuron not found in any other cortical region. Spindle cells, or Von Economo neurons (VENs), are large cells with distinct morphology (Seeley et al., 2012) that have only been found to exist in the brains of humans (Nimchinsky et al., 1999), great apes, and a select few other species (Butti, Sherwood, Hakeem, Allman, & Hof, 2009). The fact that VENs appear to be unique, from both phylogenetic and ontogenetic standpoints, has led to some interesting speculation regarding their function. For example, it has been posited that the function of these cells is to relay outputs of the AI and ACC to association cortices to aid rapid intuitive assessments of complex situations, as is necessary during social cognitive processes (Allman, Watson, Tetreault, & Hakeem, 2005).

As network neuroscience is a rapidly evolving field, we have yet to reach consensus on questions such as: How many brain networks are there? How stable and reproducible are brain networks? How do brain networks interact? and How do they change over the course of development? One of the key nodes of the salience network, the AI, appears to also participate in networks that go by other names. For example, a cinguloopercular network comprising the AI, anterior prefrontal cortex, dorsal ACC, and thalamus has been posited to perform set-maintenance activities to maintain control in the service of task goals (Dosenbach, Fair, Cohen, Schlaggar, & Petersen, 2008). Clearly the salience and cinguloopercular networks overlap anatomically, and it has not yet been resolved whether they constitute separate or entities or are merely different descriptions of the same network. The AI also participates in the ventral attention network comprising this region along with the right temporoparietal junction, middle frontal gyrus, and ventral frontal cortex. The ventral attention network is thought to mediate stimulus-driven control of attention, detecting salient environmental events and reorienting (Corbetta, Patel, & Shulman, 2008). Again, the ventral attention network appears to overlap to some degree both anatomically and functionally with the salience network. However, the ventral attention network is right lateralized, whereas the salience network is bilateral. An important direction for future research will be to characterize the extent to which each of the networks discussed here represent unique entities. This will have implications for understanding their relative functional independence and potential interactions among them.

REFERENCES

Allman, J. M., Watson, K. K., Tetreault, N. A., & Hakeem, A. Y. (2005). Intuition and autism: A possible role for Von Economo neurons. *Trends in Cognitive Sciences, 9*(8), 367–373.

Augustine, J. R. (1996). Circuitry and functional aspects of the insular lobe in primates including humans. *Brain Research Reviews, 22*(3), 229–244.

Barrett, L. F., & Satpute, A. B. (2013). Large-scale brain networks in affective and social neuroscience: Towards an integrative functional architecture of the brain. *Current Opinion in Neurobiology, 23*(3), 361–372.

Beckmann, M., Johansen-Berg, H., & Rushworth, M. F. (2009). Connectivity-based parcellation of human cingulate cortex and its relation to functional specialization. *The Journal of Neuroscience, 29*(4), 1175–1190.

Biswal, B., Yetkin, F. Z., Haughton, V. M., & Hyde, J. S. (1995). Functional connectivity in the motor cortex of resting human brain using echo-planar MRI. *Magnetic Resonance in Medicine: Official Journal of the Society of Magnetic Resonance in Medicine/Society of Magnetic Resonance in Medicine, 34*(4), 537–541.

Bressler, S. L., & Menon, V. (2010). Large-scale brain networks in cognition: Emerging methods and principles. *Trends in Cognitive Sciences, 14*(6), 277–290.

Buckner, R. L., & Carroll, D. C. (2007). Self-projection and the brain. *Trends in Cognitive Sciences, 11*(2), 49–57.

Butti, C., Sherwood, C. C., Hakeem, A. Y., Allman, J. M., & Hof, P. R. (2009). Total number and volume of Von Economo neurons in the cerebral cortex of cetaceans. *The Journal of Comparative Neurology, 515*(2), 243–259.

Corbetta, M., Patel, G., & Shulman, G. L. (2008). The reorienting system of the human brain: From environment to theory of mind. *Neuron, 58*(3), 306–324.

Damoiseaux, J. S., Rombouts, S. A., Barkhof, F., Scheltens, P., Stam, C. J., Smith, S. M., et al. (2006). Consistent resting-state networks across healthy subjects. *Proceedings of the National Academy of Sciences of the United States of America, 103*(37), 13848–13853.

Dosenbach, N. U., Fair, D. A., Cohen, A. L., Schlaggar, B. L., & Petersen, S. E. (2008). A dual-networks architecture of top-down control. *Trends in Cognitive Sciences, 12*(3), 99–105.

Fox, M. D., Corbetta, M., Snyder, A. Z., Vincent, J. L., & Raichle, M. E. (2006). Spontaneous neuronal activity distinguishes human dorsal and ventral attention systems. *Proceedings of the National Academy of Sciences of the United States of America, 103*(26), 10046–10051.

Friston, K. (1994). Functional and effective connectivity in neuroimaging: A synthesis. *Human Brain Mapping, 2*, 56–78.

Greicius, M. D., Krasnow, B., Reiss, A. L., & Menon, V. (2003). Functional connectivity in the resting brain: A network analysis of the default mode hypothesis. *Proceedings of the National Academy of Sciences of the United States of America, 100*(1), 253–258.

Mufson, E. J., & Mesulam, M. M. (1982). Insula of the old world monkey. II: Afferent cortical input and comments on the claustrum. *The Journal of Comparative Neurology, 212*(1), 23–37.

Nimchinsky, E. A., Gilissen, E., Allman, J. M., Perl, D. P., Erwin, J. M., & Hof, P. R. (1999). A neuronal morphologic type unique to humans and great apes. *Proceedings of the National Academy of Sciences of the United States of America, 96*(9), 5268–5273.

Paus, T., Castro-Alamancos, M. A., & Petrides, M. (2001). Cortico-cortical connectivity of the human mid-dorsolateral frontal cortex and its modulation by repetitive transcranial magnetic stimulation. *The European Journal of Neuroscience, 14*(8), 1405–1411.

Pessoa, L. (2014). Understanding brain networks and brain organization. *Physics of Life Reviews, 11*(3), 400–435.

Raichle, M. E., MacLeod, A. M., Snyder, A. Z., Powers, W. J., Gusnard, D. A., & Shulman, G. L. (2001). A default mode of brain function. *Proceedings of the National Academy of Sciences of the United States of America, 98*(2), 676–682.

Seeley, W. W., Menon, V., Schatzberg, A. F., Keller, J., Glover, G. H., Kenna, H., et al. (2007). Dissociable intrinsic connectivity networks for salience processing and executive control. *The Journal of Neuroscience, 27*(9), 2349–2356.

Seeley, W. W., Merkle, F. T., Gaus, S. E., Craig, A. D., Allman, J. M., & Hof, P. R. (2012). Distinctive neurons of the anterior cingulate and frontoinsular cortex: A historical perspective. *Cerebral Cortex, 22*(2), 245–250.

Shehzad, Z., Kelly, A. M., Reiss, P. T., Gee, D. G., Gotimer, K., Uddin, L. Q., et al. (2009). The resting brain: Unconstrained yet reliable. *Cerebral Cortex, 19*(10), 2209–2229.

Smith, S. M., Fox, P. T., Miller, K. L., Glahn, D. C., Fox, P. M., Mackay, C. E., et al. (2009). Correspondence of the brain's functional architecture during activation and rest. *Proceedings of the National Academy of Sciences of the United States of America, 106*(31), 13040–13045.

Taylor, K. S., Seminowicz, D. A., & Davis, K. D. (2009). Two systems of resting state connectivity between the insula and cingulate cortex. *Human Brain Mapping, 30*(9), 2731–2745.

Uddin, L. Q. (2015). Salience processing and insular cortical function and dysfunction. *Nature Reviews. Neuroscience, 16*(1), 55–61.

Uddin, L. Q., Iacoboni, M., Lange, C., & Keenan, J. P. (2007). The self and social cognition: The role of cortical midline structures and mirror neurons. *Trends in Cognitive Sciences, 11*(4), 153–157.

CHAPTER 3

Functions of the Salience Network

In a tongue-in-cheek article by Behrens and colleagues, the authors reported results from a study in which they determined which were the most "popular" brain regions based on activation likelihood. The authors found that the anterior insula was one of the most frequently reported regions of activation in studies documented in the BrainMap database (https://www.brainmap.org/). Further, a higher journal impact factor predicted activity in the anterior insula, meaning that neuroimaging studies that reported activation in this region were published in high-impact journals (Behrens, Fox, Laird, & Smith, 2013). This fun analysis highlights not only the ubiquity of insula activations in the cognitive neuroscience literature but also the relative importance researchers attach to these activations.

In the original article describing the salience network, the authors proposed that its function might be to identify the most homeostatically relevant among multiple internal and external stimuli that constantly accost the brain (Seeley et al., 2007). Prior to this work, the salience and central executive networks (CENs) were together considered to comprise a "task-positive network" or "task-activation ensemble." Seeley and colleagues demonstrated two important points in their seminal work: (1) two dissociable networks (SN and CEN) critical for guidance of behaviors can be identified by both data-driven and hypothesis-driven functional connectivity approaches, and (2) intrinsic connectivity of these two networks differentially correlates with individual differences in anxiety and executive task performance.

The salience network responds to novel stimuli across modalities. The classic "oddball" paradigm in psychology, in which participants view a series of the same item interspersed with a different item at unpredictable intervals, robustly activates the insula and dorsal anterior cingulate cortex (dACC) (Kim, 2014; Levy & Wagner, 2011) (Fig. 3). The first study to identify the anatomical "driver" within the salience network used chronometric techniques and Granger causal analyses

Salience Network of the Human Brain. DOI: http://dx.doi.org/10.1016/B978-0-12-804593-0.00003-5

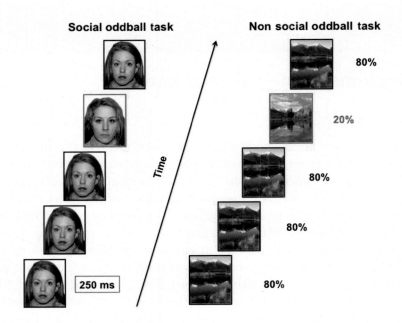

Figure 3 **The oddball paradigm.** *In classic oddball tasks, participants view a series of the same item interspersed with a different item at unpredictable intervals (Odriozola et al., 2016). A contrast examining brain regions that show greater activation in response to infrequent trials than to frequent trials typically reveals activation in the insula and dACC. Oddball tasks can use any class of visual or auditory stimuli.*

(GCA) to demonstrate that across auditory tasks, visual tasks, and task-free conditions, the right dorsal anterior insula (dAI) appears to act as a "causal outflow hub" at the junction of other large-scale neuro-cognitive networks. Specifically, the right dAI is thought to generate control signals that causally influence the default mode network (DMN, associated with self-oriented and social cognition) and central executive network (CEN, involved in the maintenance and manipulation of information and decision-making) (Sridharan, Levitin, & Menon, 2008). An intracranial EEG study examining causal interactions between brain regions during a stop-signal task demonstrated that errors were immediately followed by a feedforward influence from AI onto anterior cingulate cortex and, subsequently, onto the presupplementary motor area (Bastin et al., 2016). This finding corroborates the notion that signals from the AI serve to detect and convey error signals, making this region a likely candidate for the input to the brain's error-monitoring system. As errors tend to be salient, this fits within the broader purported function of the salience network. Dynamic causal modeling also demonstrates that the right AI drives the salience network after errors to facilitate behavioral adaptation (Ham, Leff, de Boissezon, Joffe, & Sharp, 2013). Multivariate dynamical systems approaches that overcome some of the limitations of GCA (Ryali, Supekar, Chen, & Menon, 2011) have shown that across a variety of cognitive control tasks, the AI causally influences the dACC, with the strength of causal interaction demonstrably greater on tasks requiring more cognitive control (Cai et al., 2016). Other work using graph theory—based analyses has shown increased integration between the salience network and both the CEN and DMN during high working memory loads (Liang, Zou, He, & Yang, 2016). This empirical work has been summarized in several reviews that detail the specifics of how the salience network integrates external sensory information with internal emotional and bodily state signals to coordinate brain network dynamics; namely, to initiate switches between the DMN and CEN (Menon & Uddin, 2010; Uddin, 2015).

In recent studies carefully describing the functional neuroanatomy of the insular cortex, at least three distinct subdivisions have been described. A dAI region with connections to frontal, anterior cingulate, and parietal areas is involved in cognitive control processes, a ventral anterior insula (vAI) subdivision has connections with limbic areas and is involved in affective processes, and a mid-posterior insula (PI) subdivision has connections with brain areas involved in sensorimotor processing (Chang, Yarkoni, Khaw, & Sanfey, 2013; Deen, Pitskel, & Pelphrey, 2011; Uddin, Kinnison, Pessoa, & Anderson, 2014). The salience network appears to be more tightly linked with the dAI subdivision (Uddin, Supekar, Ryali, & Menon, 2011). Using dynamic functional network connectivity analyses to examine time-varying properties of interactions between insular subdivisions and other brain regions, we recently demonstrated that the dAI exhibits more variable connections than the other insular subdivisions, and suggested that this variability may contribute to functional flexibility of the dAI (Nomi et al., 2016). This is in line with our earlier work demonstrating "diversity" of the dAI, which is active across multiple task domains (Uddin et al., 2014).

Perhaps one of the most salient experiences is that of being in pain. Painful sensation reliably activates the insula and ACC, so much so that these regions are often referred to as part of a "pain matrix." As machine learning can successfully discriminate painful from nonpainful stimuli based on fMRI signals from the insula and ACC, these regions are thought to contribute to a "neurologic signature of physical pain" (Wager et al., 2013). Although pain is such a salient stimulus that it is difficult to ignore, we have all had the experience that pain waxes and wanes as a function of our attention. This phenomenon is captured in the concept of a "dynamic pain connectome," a spatiotemporal signature of brain network communication that represents the integration of all cognitive, affective, and sensorimotor aspects of pain (Kucyi & Davis, 2015). Pain is behaviorally relevant, as it typically signals that tissue damage is occurring and something needs to be done to minimize further damage. Recently, pain-responsive brains systems have been recast as systems involved in detecting, orienting attention towards, and reacting to salient sensory stimuli, regardless of their modality (Legrain, Iannetti, Plaghki, & Mouraux, 2011). We take the view here that salient signals, whether they come from the experience of pain or some other noteworthy event, activate the cingulate and insula-based salience network.

An eye-catching study conducted in the early 2000s brought other functions of the cingulate and insular cortices into the spotlight. In a study of empathy by Singer and colleagues, the authors demonstrated that while individuals watched their partner receive a painful stimulus, the bilateral anterior insula and ACC were activated, and that these activations overlapped with those observed when the individuals themselves experienced a painful stimulus (Singer et al., 2004). Thus both nociceptive and empathic pain overlap partially in insula and ACC (Zaki, Wager, Singer, Keysers, & Gazzola, 2016).

As mentioned in Chapter 2, Anatomy of the Salience Network, a set of brain regions described as the cinguloopercular network containing overlapping nodes with the salience network has been described as a system involved in set-maintenance activities (Dosenbach, Fair, Cohen, Schlaggar, & Petersen, 2008). By contrast, our own model of anterior insular function posits a transient role for the anterior insula in detection of salient stimuli and initiation of control signals that are sustained in prefrontal cortices (Menon & Uddin, 2010). An fMRI studyusing a conflict-adaptation paradigm directly tested these competing hypotheses and found that while increased demands on moment-to-moment adjustments were associated with phasic activity in the ACC and anterior insula, increased demands on stable task-set maintenance were associated with increased sustained activity in medial superior frontal gyrus (Wilk, Ezekiel, & Morton, 2012). These findings are in line with the conceptualization of the anterior insula as part of a salience network for rapid and transient transmission of relevant information.

The concept of salience provides an umbrella under which to understand phenomena as disparate as pain, empathy, and error detection. Any event that reorients the individual and redirects his/her attention is salient, whether it affects the individual physically (as in pain) or psychologically (as with empathy). As we have all experienced, seeing someone we know suffer can be just as disturbing as suffering ourselves. The salience network appears poised to respond to situations requiring a reassessment of goals, regardless of the source and type of novel information to be processed.

REFERENCES

Bastin, J., Deman, P., David, O., Gueguen, M., Benis, D., Minotti, L., et al. (2016). Direct recordings from human anterior insula reveal its leading role within the error-monitoring network. *Cerebral Cortex* (in press).

Behrens, T. E., Fox, P., Laird, A., & Smith, S. M. (2013). What is the most interesting part of the brain? *Trends in Cognitive Sciences, 17*(1), 2−4.

Cai, W., Chen, T., Ryali, S., Kochalka, J., Li, C. S., & Menon, V. (2016). Causal interactions within a frontal-cingulate-parietal network during cognitive control: Convergent evidence from a multisite-multitask investigation. *Cerebral Cortex, 26*(5), 2140−2153.

Chang, L. J., Yarkoni, T., Khaw, M. W., & Sanfey, A. G. (2013). Decoding the role of the insula in human cognition: Functional parcellation and large-scale reverse inference. *Cerebral Cortex, 23*(3), 739−749.

Deen, B., Pitskel, N. B., & Pelphrey, K. A. (2011). Three systems of insular functional connectivity identified with cluster analysis. *Cerebral Cortex, 21*(7), 1498−1506.

Dosenbach, N. U., Fair, D. A., Cohen, A. L., Schlaggar, B. L., & Petersen, S. E. (2008). A dual-networks architecture of top-down control. *Trends in Cognitive Sciences, 12*(3), 99−105.

Ham, T., Leff, A., de Boissezon, X., Joffe, A., & Sharp, D. J. (2013). Cognitive control and the salience network: An investigation of error processing and effective connectivity. *The Journal of Neuroscience, 33*(16), 7091−7098.

Kim, H. (2014). Involvement of the dorsal and ventral attention networks in oddball stimulus processing: A meta-analysis. *Human Brain Mapping, 35*(5), 2265−2284.

Kucyi, A., & Davis, K. D. (2015). The dynamic pain connectome. *Trends in Neurosciences, 38*(2), 86−95.

Legrain, V., Iannetti, G. D., Plaghki, L., & Mouraux, A. (2011). The pain matrix reloaded: A salience detection system for the body. *Progress in Neurobiology, 93*(1), 111−124.

Levy, B. J., & Wagner, A. D. (2011). Cognitive control and right ventrolateral prefrontal cortex: Reflexive reorienting, motor inhibition, and action updating. *Annals of the New York Academy of Sciences, 1224*, 40−62.

Liang, X., Zou, Q., He, Y., & Yang, Y. (2016). Topologically reorganized connectivity architecture of default-mode, executive-control, and salience networks across working memory task loads. *Cerebral Cortex, 26*(4), 1501−1511.

Menon, V., & Uddin, L. Q. (2010). Saliency, switching, attention and control: A network model of insula function. *Brain Structure & Function, 214*(5−6), 655−667.

Nomi, J. S., Farrant, K., Damaraju, E., Rachakonda, S., Calhoun, V. D., & Uddin, L. Q. (2016). Dynamic functional network connectivity reveals unique and overlapping profiles of insula subdivisions. *Human Brain Mapping, 37*(5), 1770−1787.

Odriozola, A., Uddin, L. Q., Lynch, C. J., Kochalka, J., Chen, T., & Menon, V. (2016). Insula response and connectivity during social and non-social attention in children with autism. *Social Cognitive and Affective Neuroscience, 11*(3), 433−444.

Ryali, S., Supekar, K., Chen, T., & Menon, V. (2011). Multivariate dynamical systems models for estimating causal interactions in fMRI. *Neuroimage, 54*(2), 807−823.

Seeley, W. W., Menon, V., Schatzberg, A. F., Keller, J., Glover, G. H., Kenna, H., et al. (2007). Dissociable intrinsic connectivity networks for salience processing and executive control. *The Journal of Neuroscience, 27*(9), 2349−2356.

Singer, T., Seymour, B., O'Doherty, J., Kaube, H., Dolan, R. J., & Frith, C. D. (2004). Empathy for pain involves the affective but not sensory components of pain. *Science (New York, N.Y.), 303*(5661), 1157−1162.

Sridharan, D., Levitin, D. J., & Menon, V. (2008). A critical role for the right fronto-insular cortex in switching between central-executive and default-mode networks. *Proceedings of the National Academy of Sciences of the United States of America, 105*(34), 12569−12574.

Uddin, L. Q. (2015). Salience processing and insular cortical function and dysfunction. *Nature Reviews. Neuroscience, 16*(1), 55−61.

Uddin, L. Q., Kinnison, J., Pessoa, L., & Anderson, M. L. (2014). Beyond the tripartite cognition-emotion-interoception model of the human insular cortex. *Journal of Cognitive Neuroscience, 26*(1), 16−27.

Uddin, L. Q., Supekar, K. S., Ryali, S., & Menon, V. (2011). Dynamic reconfiguration of structural and functional connectivity across core neurocognitive brain networks with development. *The Journal of Neuroscience, 31*(50), 18578−18589.

Wager, T. D., Atlas, L. Y., Lindquist, M. A., Roy, M., Woo, C. W., & Kross, E. (2013). An fMRI-based neurologic signature of physical pain. *The New England Journal of Medicine, 368*(15), 1388−1397.

Wilk, H. A., Ezekiel, F., & Morton, J. B. (2012). Brain regions associated with moment-to-moment adjustments in control and stable task-set maintenance. *Neuroimage, 59*(2), 1960−1967.

Zaki, J., Wager, T. D., Singer, T., Keysers, C., & Gazzola, V. (2016). The anatomy of suffering: Understanding the relationship between nociceptive and empathic pain. *Trends in Cognitive Sciences, 20*(4), 249−259.

CHAPTER 4

Salience Network Across the Life Span

Now that we have spent some time reviewing the anatomy and functions of the salience network, let us turn to the question of how this network emerges throughout development, changes across the life span, and continues to support flexible behaviors into old age.

Brain networks involved in high-level cognitive processes such as those relying on the salience network emerge across early development. In fetuses 24—38 weeks of gestation, brain networks that comprise primary motor and visual areas can be detected using independent component analysis applied to resting state functional magnetic resonance imaging (fMRI) data (Thomason et al., 2013), but the salience network does not appear to be detectable until after birth. Newborn infants have cortical hubs in mostly motor, sensory, and visual cortices, whereas adults have cortical hubs in association cortices including medial prefrontal cortex and the insula (Fransson, Aden, Blennow, & Lagercrantz, 2011). Other longitudinal work characterizing the emergence of brain networks during the first 2 years of life has also noted that primary networks are topologically adult-like in neonates, whereas higher-order networks are topologically incomplete and isolated. The salience network shows age-dependent decreases in functional connectivity (e.g., segregation from other networks) between birth and 2 years of age (Gao, Alcauter, Smith, Gilmore, & Lin, 2015). Infants born very preterm (e.g., at or before gestational age of 33 weeks) who were tested in adulthood show reduced functional connectivity of the insula compared with term-born individuals (White et al., 2014), suggesting that brain network abnormalities that occur as a result of preterm birth can persist well into adulthood.

Salience network development across childhood and adolescence has received quite a bit of attention in recent years. Comparing children (age 7—9 years) with young adults (age 19—22 years), we found greater functional and effective connectivity between the right anterior insula and ACC and DLPFC with development. These changes in

Salience Network of the Human Brain. DOI: http://dx.doi.org/10.1016/B978-0-12-804593-0.00004-7

functional coupling were accompanied by observed increases in structural connectivity along the uncinate and fronto-occipital fasciculi between childhood and adulthood (Uddin, Supekar, Ryali, & Menon, 2011) (Fig. 4). Earlier work also hinted at this pattern of strengthening of rFIC-ACC connections with age (Fair et al., 2007), though more recent studies have demonstrated mixed findings (including patterns of both increased and decreased connectivity) regarding salience network changes with development (Sole-Padulles et al., 2016).

As reviewed in Chapter 3, Functions of the Salience Network, one of the functions of the salience network is task control, or the ability to select appropriate behaviors in line with current goals. One cognitive interference task designed to assess task control is the Multisource Interference Task (MSIT), which requires participants to press a button corresponding to the unique number in a string of three digits (e.g., "331"). The task becomes difficult (high interference) when the position

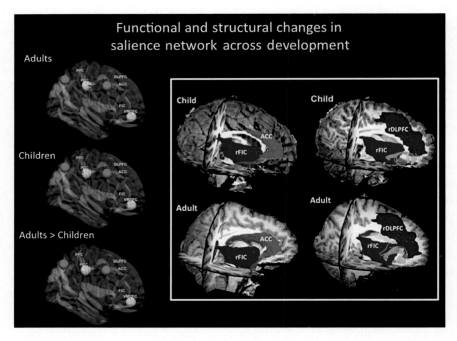

Figure 4 **Functional and structural changes in the salience network across development.** *Greater effective connectivity between the right anterior insula and ACC and DLPFC can be seen in typical development. These changes in functional coupling are paralleled by increases in structural connectivity along the uncinate and fronto-occipital fasciculi between childhood and adulthood (Uddin, Supekar, Ryali, & Menon, 2011).*

of the target is incongruent with its associated response (e.g., for "331", "1" is in the third position, but requires a button press with the first, or index, finger). A study of brain activation and connectivity in adolescents between the age 8 and 19 years performing the MSIT demonstrated that age-related increases in functional connectivity can be observed between preSMA regions important for task performance and the anterior insular cortex. This study also found nonlinear developmental patterns of activation (reduced in middle adolescence compared with younger and older ages) for the dorsal ACC (Liu, Angstadt, Taylor, & Fitzgerald, 2016). These changes in activation and connectivity patterns of the salience network are thought to underlie improvements in task control across development. Another fMRI study assessing changes in the ability to implement rapid adjustments in control between the age 9 and 32 years demonstrated that the anterior insula and anterior cingulate were associated with moment-to-moment adjustment in older but not younger participants. These findings confirm the importance of the salience network in control processes, and demonstrate continuous increases in the activation of this system across development (Wilk & Morton, 2012).

Very few studies have attempted a comprehensive analysis of connectivity of the salience network over the life span. In a recent study examining relationships between the SN, default mode network (DMN), and central executive network (CEN) in individuals between the age 21 and 79 years, Archer and colleagues collected fMRI data during resting states and visuospatial working memory and inhibition tasks. They found that connectivity decreased with age in most networks both during the resting state and task performance. In the resting state, they found increased between-network connectivity (SN-CEN) with older age. For working memory and inhibition tasks, increased connectivity was also observed between SN and DMN and CEN with increasing age (Archer, Lee, Qiu, & Chen, 2016). This study demonstrates the ways in which SN connectivity changes across the life span to integrate more closely with other large-scale brain networks.

In late life, dynamic coordination of large-scale brain networks is affected in ways that are only now beginning to be understood. Comparing older participants (age 59–81 years) with younger individuals (age 23–37 years) reveals reduced functional connectivity of the SN in older adults (Marstaller, Williams, Rich, Savage, & Burianova, 2015).

Successful cognitive aging requires preservation of interactions within and between large-scale brain networks across the life span. In a study of the SN, DMN, and dorsal attention network in adults between the age 18 and 88 years, Tsvetanov and colleagues demonstrated an age-dependent effect of brain network connectivity on cognitive function such that cognitive performance relied on neural dynamics more strongly in older adults (Tsvetanov et al., 2016). Focusing on the SN, they noted that age was independently predicted by decreased effective connectivity between dACC and left AI and increased effective connectivity from right AI to left AI. Another recent study of 186 older adults between the age 65 and 85 years found a loss of connect-edness of the right anterior insula in older participants (Muller, Merillat, & Jancke, 2016). These findings contribute to a growing body of literature proposing links between brain dynamics and success-ful aging. In Chapter 5, Salience Network Dysfunction, we will delve into the role of reduced SN integrity in psychiatric and neurological conditions.

REFERENCES

Archer, J. A., Lee, A., Qiu, A., & Chen, S. H. (2016). A comprehensive analysis of connectivity and aging over the adult life span. *Brain Connectivity, 6*(2), 169−185.

Fair, D. A., Dosenbach, N. U., Church, J. A., Cohen, A. L., Brahmbhatt, S., Miezin, F. M., et al. (2007). Development of distinct control networks through segregation and integration. *Proceedings of the National Academy of Sciences of the United States of America, 104*(33), 13507−13512.

Fransson, P., Aden, U., Blennow, M., & Lagercrantz, H. (2011). The functional architecture of the infant brain as revealed by resting-state fMRI. *Cerebral Cortex, 21*(1), 145−154.

Gao, W., Alcauter, S., Smith, J. K., Gilmore, J. H., & Lin, W. (2015). Development of human brain cortical network architecture during infancy. *Brain Structure & Function, 220*(2), 1173−1186.

Liu, Y., Angstadt, M., Taylor, S. F., & Fitzgerald, K. D. (2016). The typical development of posterior medial frontal cortex function and connectivity during task control demands in youth 8-19 years old. *Neuroimage, 137*, 97−106.

Marstaller, L., Williams, M., Rich, A., Savage, G., & Burianova, H. (2015). Aging and large-scale functional networks: White matter integrity, gray matter volume, and functional connectivity in the resting state. *Neuroscience, 290*, 369−378.

Muller, A. M., Merillat, S., & Jancke, L. (2016). Small changes, but huge impact? The right anterior insula's loss of connection strength during the transition of old to very old age. *Frontiers in Aging Neuroscience, 8*, 86.

Sole-Padulles, C., Castro-Fornieles, J., de la Serna, E., Calvo, R., Baeza, I., Moya, J., et al. (2016). Intrinsic connectivity networks from childhood to late adolescence: Effects of age and sex. *Development in Cognitive Neuroscience, 17*, 35−44.

Thomason, M. E., Dassanayake, M. T., Shen, S., Katkuri, Y., Alexis, M., Anderson, A. L., et al. (2013). Cross-hemispheric functional connectivity in the human fetal brain. *Science Translation Medicine, 5*(173), 173ra124.

Tsvetanov, K. A., Henson, R. N., Tyler, L. K., Razi, A., Geerligs, L., Ham, T. E., et al. (2016). Extrinsic and intrinsic brain network connectivity maintains cognition across the lifespan despite accelerated decay of regional brain activation. *The Journal of Neuroscience, 36*(11), 3115−3126.

Uddin, L. Q., Supekar, K. S., Ryali, S., & Menon, V. (2011). Dynamic reconfiguration of structural and functional connectivity across core neurocognitive brain networks with development. *The Journal of Neuroscience, 31*(50), 18578−18589.

White, T. P., Symington, I., Castellanos, N. P., Brittain, P. J., Froudist Walsh, S., Nam, K. W., et al. (2014). Dysconnectivity of neurocognitive networks at rest in very-preterm born adults. *Neuroimage. Clinical, 4*, 352−365.

Wilk, H. A., & Morton, J. B. (2012). Developmental changes in patterns of brain activity associated with moment-to-moment adjustments in control. *Neuroimage, 63*(1), 475−484.

CHAPTER 5

Salience Network Dysfunction

What happens when the salience network is dysfunctional? One can envision widespread consequences of aberrant salience detection. Studies examining the structural integrity, functional activation, and connectivity of the insula and dACC across neurological and psychiatric disorders are beginning to suggest that there may be clues to a variety of psychopathologies within these key brain regions.

Even though conventional wisdom suggests that individual brain disorders result from abnormalities of specific brain regions, recent work suggests common neurobiological substrates may contribute to a host of mental illnesses. In a meta-analysis of 193 voxel-based morphometry studies of individuals with schizophrenia, bipolar disorder, depression, addiction, obsessive-compulsive disorder, and anxiety, Goodkind and colleagues found convergent patterns of gray matter loss across diagnoses in dACC and bilateral insular cortices, the very regions we have been describing as part of the salience network (Goodkind et al., 2015).

Our own work has focused on autism spectrum disorder (ASD), a neurodevelopmental disorder that affects 1 in 68 children and is characterized by diminished attention to social stimuli such as human faces (Carver & Dawson, 2002). Meta-analyses of fMRI studies using social cognitive tasks (Di Martino et al., 2009) and task-related functional connectivity analyses (Odriozola et al., 2016) point to altered insula activation and connectivity in individuals with ASD (Uddin & Menon, 2009). In children with ASD aged 7–12 years old, we find hyperconnectivity of the salience network that is linked with severity of restricted and repetitive behaviors (Uddin et al., 2013), though others have reported underconnectivity of the network in relation to ASD symptomatology (Abbott et al., 2015) as well as mixed patterns of over- and underconnectivity of the salience network (Elton, Di Martino, Hazlett, & Gao, 2015). Patterns of functional connectivity of the insular cortex (Anderson et al., 2011) and salience network

Salience Network of the Human Brain. DOI: http://dx.doi.org/10.1016/B978-0-12-804593-0.00005-9

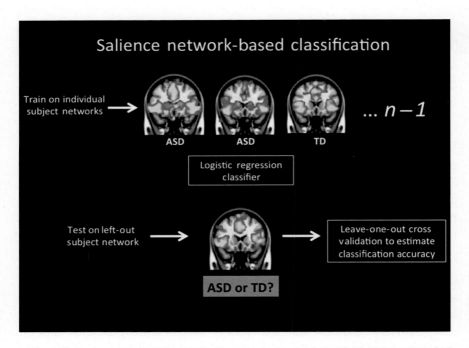

*Figure 5 **Salience network-based classification of autism.** The salience network identified from each individual subject can serve as a feature to be input into classification analyses. An algorithm is trained on a subset of the neuro-imaging data and tested on another subset of data. Signals from the salience network can be used to identify individuals with autism with high accuracy (Uddin et al., 2013).*

(Uddin et al., 2013) can be used to discriminate individuals with ASD from typically developing individuals, suggesting that signals within this network may be used to derive informative brain-based biomarkers (Fig. 5). The severity of ASD symptoms also correlates with patterns of functional connections among the dorsal anterior insula and nodes of the DMN and CEN (Uddin et al., 2015).

Another disorder with onset relatively early in life (e.g., adolescence or young adulthood) is schizophrenia, characterized by delusions, hallucinations, and flat affect (Andreasen, 2000). These symptoms, psychosis in particular, are thought to arise from a dysregulated hyperdopaminergic brain state resulting in aberrant assignment of salience to experiences. The idea is that aberrant salience detection can result in delusions (from cognitive efforts to make sense of aberrant salient experiences) and hallucinations (direct experience of the aberrant salience of internal representations) (Kapur, 2003). Patients with schizophrenia consistently exhibit reduced volume of bilateral insular cortical gray matter (Glahn et al., 2008), and reduced activation of the insula during emotion-regulation tasks can be observed in individuals with the disorder (Verbiest et al., 2014). One hypothesis is that insular dysfunction may diminish the capacity for individuals with schizophrenia to discriminate between self-generated and external information (Wylie & Tregellas, 2010). In addition to volumetric and functional activation alterations of the insula in schizophrenia, individuals with the disorder exhibit reduced strength of functional connectivity of this brain region (Wang et al., 2014). Several recent studies analyzed effective connectivity between brain regions in schizophrenia and found evidence for reduced causal influence from the anterior insula to the CEN and DMN (Manoliu et al., 2014; Moran et al., 2013; Palaniyappan, Simmonite, White, Liddle, & Liddle, 2013). Based on these lines of evidence, some have proposed that the salience network may be an appropriate therapeutic target for individuals with schizophrenia (Palaniyappan, White, & Liddle, 2012).

Prevalent internalizing disorders such as anxiety and depression are also increasingly tied to salience network dysfunction. Hyperactivation of the insula in response to fear-eliciting stimuli is often observed in individuals with a range of anxiety disorders (Peterson, Thome, Frewen, & Lanius, 2014). A meta-analysis of fMRI studies demonstrated that relative to healthy controls, individuals with major depression showed greater response in the insula and dACC when processing negative stimuli. The authors suggest that heightened activation of the salience network in response to negative information underlies symptoms of depression (Hamilton et al., 2012).

Brain health later in life also appears to rely on salience network integrity. Frontotemporal dementia (FTD) is one disorder that emerges in the sixth decade of life and is characterized by social-emotional dysfunction and aphasia. Degeneration of the frontoinsular cortices and pregenual ACC spreading to adjacent frontal regions occurs in FTD, as the name suggests (Seeley, Crawford, Zhou, Miller, & Greicius, 2009; Seeley, Zhou, & Kim, 2012). Postmortem quantitative studies demonstrate that von Economo neurons are selectively vulnerable in FTD (Kim et al., 2012). Neuroimaging meta-analyses identify the right anterior insular cortex as the most consistently affected structure in individuals with FTD (Schroeter, Raczka, Neumann, & von Cramon, 2008), and the disorder is associated with reduced intrinsic functional connectivity of the salience network (Zhou et al., 2010). As the salience network changes across the course of FTD progression, signals from this network can predict disease progression as reflected by increases in symptom severity (Day et al., 2013).

Patients with Alzheimer's disease (AD), another dementia with late-life onset, also show alterations in salience network integrity. AD patients exhibit increased resting state functional connectivity within the salience network. Further, in these patients one can find a correlation between increased connectivity in anterior cingulate cortex and right insula areas of the SN and hyperactivity syndrome (agitation, irritability, aberrant motor behavior, euphoria, and disinhibition) (Balthazar et al., 2014).

Dysfunction of different portions of the SN at various time points across the life span can produce different cognitive and behavioral phenotypes. Understanding the makeup, development, and decline of this network is the first step toward identifying ways to ensure its healthy functioning.

REFERENCES

Abbott, A. E., Nair, A., Keown, C. L., Datko, M., Jahedi, A., Fishman, I., et al. (2015). Patterns of atypical functional connectivity and behavioral links in autism differ between default, salience, and executive networks. *Cerebral Cortex*.

Anderson, J. S., Nielsen, J. A., Froehlich, A. L., DuBray, M. B., Druzgal, T. J., Cariello, A. N., et al. (2011). Functional connectivity magnetic resonance imaging classification of autism. *Brain*, *134*(Pt. 12), 3742–3754.

Andreasen, N. C. (2000). Schizophrenia: The fundamental questions. *Brain Research Reviews*, *31*(2–3), 106–112.

Balthazar, M. L., Pereira, F. R., Lopes, T. M., da Silva, E. L., Coan, A. C., Campos, B. M., et al. (2014). Neuropsychiatric symptoms in Alzheimer's disease are related to functional connectivity alterations in the salience network. *Human Brain Mapping*, *35*(4), 1237–1246.

Carver, L. J., & Dawson, G. (2002). Development and neural bases of face recognition in autism. *Molecular Psychiatry*, *7*(Suppl. 2), S18–S20.

Day, G. S., Farb, N. A., Tang-Wai, D. F., Masellis, M., Black, S. E., Freedman, M., et al. (2013). Salience network resting-state activity: Prediction of frontotemporal dementia progression. *JAMA Neurology*, *70*(10), 1249–1253.

Di Martino, A., Ross, K., Uddin, L. Q., Sklar, A. B., Castellanos, F. X., & Milham, M. P. (2009). Functional brain correlates of social and nonsocial processes in autism spectrum disorders: An activation likelihood estimation meta-analysis. *Biological Psychiatry*, *65*(1), 63–74.

Elton, A., Di Martino, A., Hazlett, H. C., & Gao, W. (2015). Neural connectivity evidence for a categorical-dimensional hybrid model of autism spectrum disorder. *Biological Psychiatry*, *80*(2), 120–128.

Glahn, D. C., Laird, A. R., Ellison-Wright, I., Thelen, S. M., Robinson, J. L., Lancaster, J. L., et al. (2008). Meta-analysis of gray matter anomalies in schizophrenia: Application of anatomic likelihood estimation and network analysis. *Biological Psychiatry, 64*(9), 774−781.

Goodkind, M., Eickhoff, S. B., Oathes, D. J., Jiang, Y., Chang, A., Jones-Hagata, L. B., et al. (2015). Identification of a common neurobiological substrate for mental illness. *JAMA Psychiatry, 72*(4), 305−315.

Hamilton, J. P., Etkin, A., Furman, D. J., Lemus, M. G., Johnson, R. F., & Gotlib, I. H. (2012). Functional neuroimaging of major depressive disorder: A meta-analysis and new integration of base line activation and neural response data. *The American Journal of Psychiatry, 169*(7), 693−703.

Kapur, S. (2003). Psychosis as a state of aberrant salience: A framework linking biology, phenomenology, and pharmacology in schizophrenia. *The American Journal of Psychiatry, 160*(1), 13−23.

Kim, E. J., Sidhu, M., Gaus, S. E., Huang, E. J., Hof, P. R., Miller, B. L., et al. (2012). Selective frontoinsular von Economo neuron and fork cell loss in early behavioral variant frontotemporal dementia. *Cerebral Cortex, 22*(2), 251−259.

Manoliu, A., Riedl, V., Zherdin, A., Muhlau, M., Schwerthoffer, D., Scherr, M., et al. (2014). Aberrant dependence of default mode/central executive network interactions on anterior insular salience network activity in schizophrenia. *Schizophrenia Bulletin, 40*(2), 428−437.

Moran, L. V., Tagamets, M. A., Sampath, H., O'Donnell, A., Stein, E. A., Kochunov, P., et al. (2013). Disruption of anterior insula modulation of large-scale brain networks in schizophrenia. *Biological Psychiatry, 74*(6), 467−474.

Odriozola, P., Uddin, L. Q., Lynch, C. J., Kochalka, J., Chen, T., & Menon, V. (2016). Insula response and connectivity during social and non-social attention in children with autism. *Social Cognitive and Affective Neuroscience, 11*(3), 433−444.

Palaniyappan, L., Simmonite, M., White, T. P., Liddle, E. B., & Liddle, P. F. (2013). Neural primacy of the salience processing system in schizophrenia. *Neuron, 79*(4), 814−828.

Palaniyappan, L., White, T. P., & Liddle, P. F. (2012). The concept of salience network dysfunction in schizophrenia: From neuroimaging observations to therapeutic opportunities. *Current Topics in Medicinal Chemistry, 12*(21), 2324−2338.

Peterson, A., Thome, J., Frewen, P., & Lanius, R. A. (2014). Resting-state neuroimaging studies: A new way of identifying differences and similarities among the anxiety disorders? *Canadian Journal of Psychiatry, 59*(6), 294−300.

Schroeter, M. L., Raczka, K., Neumann, J., & von Cramon, D. Y. (2008). Neural networks in frontotemporal dementia—A meta-analysis. *Neurobiology of Aging, 29*(3), 418−426.

Seeley, W. W., Crawford, R. K., Zhou, J., Miller, B. L., & Greicius, M. D. (2009). Neurodegenerative diseases target large-scale human brain networks. *Neuron, 62*(1), 42−52.

Seeley, W. W., Zhou, J., & Kim, E. J. (2012). Frontotemporal dementia: What can the behavioral variant teach us about human brain organization? *Neuroscientist, 18*(4), 373−385.

Uddin, L. Q., & Menon, V. (2009). The anterior insula in autism: Under-connected and under-examined. *Neuroscience and Biobehavioral Reviews, 33*(8), 1198−1203.

Uddin, L. Q., Supekar, K., Lynch, C. J., Cheng, K. M., Odriozola, P., Barth, M. E., et al. (2015). Brain state differentiation and behavioral inflexibility in autism. *Cerebral Cortex, 25*(12), 4740−4747.

Uddin, L. Q., Supekar, K., Lynch, C. J., Khouzam, A., Phillips, J., Feinstein, C., et al. (2013). Salience network-based classification and prediction of symptom severity in children with autism. *JAMA Psychiatry, 70*(8), 869−879.

Verbiest, M. E., Crone, M. R., Scharloo, M., Chavannes, N. H., van der Meer, V., Kaptein, A. A., et al. (2014). One-hour training for general practitioners in reducing the implementation gap of smoking cessation care: A cluster-randomized controlled trial. *Nicotine & Tobacco Research: Official Journal of the Society for Research on Nicotine and Tobacco, 16*(1), 1−10.

Wang, X., Xia, M., Lai, Y., Dai, Z., Cao, Q., Cheng, Z., et al. (2014). Disrupted resting-state functional connectivity in minimally treated chronic schizophrenia. *Schizophrenia Research, 156* (2−3), 150−156.

Wylie, K. P., & Tregellas, J. R. (2010). The role of the insula in schizophrenia. *Schizophrenia Research, 123*(2−3), 93−104.

Zhou, J., Greicius, M. D., Gennatas, E. D., Growdon, M. E., Jang, J. Y., Rabinovici, G. D., et al. (2010). Divergent network connectivity changes in behavioural variant frontotemporal dementia and Alzheimer's disease. *Brain, 133*(Pt. 5), 1352−1367.

CHAPTER 6

Future Directions in Salience Network Research

So far we have covered the current research on salience network anatomy, function and dysfunction across the lifespan. In Chapter 3, Functions of the Salience Network, Chapter 4, Salience Network Across the Life Span, and Chapter 5, Salience Network Dysfunction, we reviewed studies of the salience network that used traditional neuroimaging approaches including functional connectivity and function coactivation analyses. Most of these traditional approaches assume that functional connections are static, or stable, over time. However, functional connections in the brain can form transiently; thus it is important to characterize the nature of these dynamics to create a more complete understanding of the salience network of the human brain.

A new approach has emerged over the last several years for assessing variations in brain network topology over time. These novel approaches are referred to as "dynamic functional connectivity" (Chang & Glover, 2010) or "dynamic functional network connectivity" (dFNC) (Allen et al., 2014), analyses, and they can be used to quantify moment-to-moment changes in functional connectivity metrics. For example, one can use a sliding window approach, whereby a time window of fixed length is selected, and data points within that window are used to calculate the functional connectivity metric of interest. The window is then shifted in time by a fixed number of data points (hence, sliding) (Fig. 6). This process results in quantification of the time-varying property of the chosen metric over the duration of data acquisition (Hutchison et al., 2013). Dynamic functional connectivity approaches have already begun to reveal unique insights into salience network organization. For example, Chang and colleagues showed that the posterior cingulate cortex, a prominent node of the default mode network, exhibits variable correlations with the insular nodes of the salience network (Chang & Glover, 2010).

As reviewed in Chapter 2, Anatomy of the Salience Network, and Chapter 3, Functions of the Salience Network, distinct subdivisions

Salience Network of the Human Brain. DOI: http://dx.doi.org/10.1016/B978-0-12-804593-0.00006-0

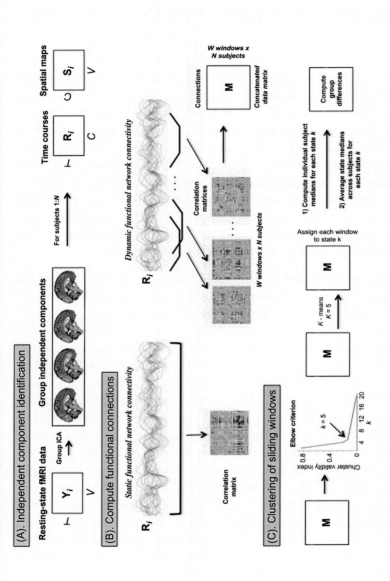

Figure 6 *Analytic procedure to compute dynamic functional connectivity.* *(A) Independent component analysis (ICA) is used to create a functional parcellation of the brain. (B) Subject-specific time courses from the group ICA are then used to calculate functional connections. The dynamic functional connectivity analysis utilizes sliding windows to compute pairwise correlations between all brain regions within a bin of specified length. (C) Clustering is applied to the resulting correlation matrices to estimate the number of underlying brain states. Group differences in connectivity dynamics can then be assessed (Allen et al., 2014; Nomi et al., 2016).*

within the insular cortex are thought to subserve partially discrete functions. We have recently found that transient connectivity profiles of insula subdivisions can be delineated using a dFNC approach, and that the dorsal anterior insula exhibits the greatest functional connectivity variability (Nomi et al., 2016). Chen and colleagues have also recently demonstrated distinct global brain dynamics that characterize the spatiotemporal organization of the salience network. They find that the salience network demonstrates the highest levels of flexibility in time-varying connectivity with other brain networks important for attention and control (Chen, Cai, Ryali, Supekar, & Menon, 2016). We are optimistic that as the field of functional connectivity dynamics grows in methodological sophistication, we will see a surge in careful descriptions of salience network dynamics.

Another future direction for salience network research involves the potential utility of deriving biomarkers from brain regions comprising this network. The term "biomarker" or "biological marker" refers to a broad category of medical signs, or objective indications of a medical state, that can be measured accurately and reproducibly and may influence and predict the incidence and outcome of disease (Strimbu & Tavel, 2010). Increasingly, clinical neuroscience has embraced the possibility that metrics derived from brain imaging may be used to predict the diagnostic category in a variety of psychiatric and neurological conditions. Accordingly, another promising future direction for salience network research is working toward the goal of using signals from this network to develop brain-based biomarkers for objective identification of specific disorders.

Using machine learning approaches, previous studies have found evidence that functional connectivity of the salience network can discriminate children with autism from typically developing children (Anderson et al., 2011; Uddin et al., 2013). Similar results have been reported for disorders with later-life onsets including frontotemporal dementia (Day et al., 2013). While salience network dysfunction seems to be a characteristic of several prevalent disorders, it is not yet clear specifically which aspects of network dysfunction contribute to which specific symptoms. Knowledge of these precise brain–behavior relationships will facilitate more targeted treatment and intervention strategies.

Salience processing often impacts memory processes and the extent to which particular events are encoded and subsequently maintained in long-term memory. At present, the impact of salience on memory in clinical populations is entirely unexplored. Many late-life disorders and dementias involve memory loss. A recent positron emission tomography study of patients with Parkinson's disease found that memory-impaired patients showed reductions in D2 receptor binding in the salience network compared with healthy controls (Christopher et al., 2015). The complex interplay between dopamine function, salience processing, and memory impairment are thus only beginning to be explored, and will likely be the topic of future research in aging and brain function.

The past 20 years have witnessed rapid advances in the use of functional MRI to enhance our understanding of human brain function and dysfunction. The science of the salience network and its interactions with other large-scale brain networks is still in its infancy. Continued investigations into the functional role of the salience network across neurotypical and clinical populations will be key to understanding how the human brain responds flexibly and adaptively to challenges.

REFERENCES

Allen, E. A., Damaraju, E., Plis, S. M., Erhardt, E. B., Eichele, T., & Calhoun, V. D. (2014). Tracking whole-brain connectivity dynamics in the resting state. *Cerebral Cortex, 24*(3), 663−676.

Anderson, J. S., Nielsen, J. A., Froehlich, A. L., DuBray, M. B., Druzgal, T. J., Cariello, A. N., et al. (2011). Functional connectivity magnetic resonance imaging classification of autism. *Brain, 134*(Pt. 12), 3742−3754.

Chang, C., & Glover, G. H. (2010). Time-frequency dynamics of resting-state brain connectivity measured with fMRI. *Neuroimage, 50*(1), 81−98.

Chen, T., Cai, W., Ryali, S., Supekar, K., & Menon, V. (2016). Distinct global brain dynamics and spatiotemporal organization of the salience network. *PLoS Biology, 14*(6), e1002469.

Christopher, L., Duff-Canning, S., Koshimori, Y., Segura, B., Boileau, I., Chen, R., et al. (2015). Salience network and parahippocampal dopamine dysfunction in memory-impaired Parkinson disease. *Annals of Neurology, 77*(2), 269−280.

Day, G. S., Farb, N. A., Tang-Wai, D. F., Masellis, M., Black, S. E., Freedman, M., et al. (2013). Salience network resting-state activity: Prediction of frontotemporal dementia progression. *JAMA Neurology, 70*(10), 1249−1253.

Hutchison, R. M., Womelsdorf, T., Allen, E. A., Bandettini, P. A., Calhoun, V. D., Corbetta, M., et al. (2013). Dynamic functional connectivity: Promise, issues, and interpretations. *Neuroimage, 80*, 360−378.

Nomi, J. S., Farrant, K., Damaraju, E., Rachakonda, S., Calhoun, V. D., & Uddin, L. Q. (2016). Dynamic functional network connectivity reveals unique and overlapping profiles of insula subdivisions. *Human Brain Mapping, 37*(5), 1770−1787.

Strimbu, K., & Tavel, J. A. (2010). What are biomarkers? *Current Opinion in HIV and AIDS, 5*(6), 463−466.

Uddin, L. Q., Supekar, K., Lynch, C. J., Khouzam, A., Phillips, J., Feinstein, C., et al. (2013). Salience network-based classification and prediction of symptom severity in children with autism. *JAMA Psychiatry, 70*(8), 869−879.

INDEX

Note: Page numbers followed by "*f*" refer to figures.

Printed in the United States
By Bookmasters